D0466487

THE LIBRARY OF SATELLITES
SPY SATELLITES
PAUL KUPPERBERG
the rosen publishing group's
rosen
central

This one's for Robin, just because

Published in 2003 by The Rosen Publishing Group, Inc.
29 East 21st Street, New York, NY 10010

First Edition

Library of Congress Cataloging-in-Publication Data

Kupperberg, Paul.
Spy satellites/Paul Kupperberg. —1st ed.
p. cm. — (The library of satellites)
Summary: Examines the history, technology, and uses of spy satellites, looking especially at the various reconnaissance satellite programs of the United States, from the mid-twentieth century to the present.
Includes bibliographical references and index.
ISBN 0-8239-3854-9 (library binding)
1. Space surveillance—United States—Juvenile literature.
[1. Space surveillance.] I. Title. II. Series.
UG1523.K87 2002
327.1273—dc21

2002010747

Manufactured in the United States of America

TABLE OF CONTENTS

INTRODUCTION

September 11, 2001: while the world watched on television as the terrorist attacks on the World Trade Center in New York and the Pentagon just outside Washington, D.C., unfolded, there was a very small audience with an even more dramatic view of events—hundreds of miles above the scenes of devastation.

"As we went over Maine, we could see New York City and the smoke from the fires," said Frank Culbertson, the crew commander of the international space station on September 11, according to CNN.com. In orbit 240 miles (386 kilometers) above Earth with two Russian crewmates, Commander Culbertson recorded the horror in New York, visible to these spectators as a vast plume of smoke rising from below, with a camcorder. The three pairs of human eyes aboard the orbiting space station were not the only observers looking down from space, however.

Nor have these other "eyes" stopped watching in the months since the September 11 attacks. Only now, instead of being passive observers to disaster, they have become active and indispensable participants in the fight against the perpetrators of these deeds. How many of these "eyes" are watching? Who are they? Where are they? What are they looking at now?

They are spy satellites. But just about everything else about them is classified "top secret," and concrete information on them is very hard to come by.

You have probably seen images taken by satellites and relayed back to Earth. During recent times of war, such as the Persian Gulf War in 1990 or Operation Enduring Freedom in Afghanistan in 2001–2002, military press briefings often featured enlarged photos of enemy bunkers or weapons installations, taken by a reconnaissance satellite orbiting far overhead (reconnaissance is the secretive survey of enemy territory). Environmental conferences on subjects like global warming or deforestation often feature images taken by Earth imaging satellites that can indicate changes in land and water temperature, the shrinking or expanding of polar ice caps, or the loss of thousands of acres of trees. The luminous photos of Mars, the rings of Saturn, and Jupiter's moons that we have seen in recent years were provided by the cameras on board deep space probes that enter a planet's or a moon's orbit, take photographs, gather data, and then exit the orbit and move on to the next subject of study. Chances are the weather report featured on your local television station includes satellite maps of the

Smoke from the smoldering World Trade Center spreads beyond lower Manhattan, trailing over the water to New Jersey following the terrorist attacks of September 11, 2001. This photograph was taken by the *Terra* satellite, which follows a polar orbit perpendicular to the direction of Earth's spin.

country that illustrate with moving pictures the approach and severity of storm systems.

Although the work of Earth imaging, weather, communications, and deep space satellites is very familiar to us, the efforts of spy satellites are shrouded in mystery and secrecy. Very few people really know just what state-of-the-art spy satellites are capable of doing. The ones who do are rarely willing to break security and speak on the record.

What is common knowledge is that the United States has a network of reconnaissance satellites of various types and functions orbiting the planet. Exactly where these spy satellites—operated by the supersecret National Reconnaissance Office (NRO)—are located and how they gather and relay data is classified information. A spy satellite program code-named CORONA provides an excellent sense of the high level of security surrounding space-based espionage. CORONA launched 145 missions between 1960 and 1972, yet it remained an absolute secret until 1995, twenty-three years after its completion.

When attempting to piece together the missions and operations of spy satellites, all we have to go on are informed guesses made by experts in the intelligence field. If the CORONA program's history is any indication, it may be years—even decades—until information about currently operating spy satellites is declassified. One thing is certain, however: Spy satellites have served the U.S. intelligence and military communities since the days of the Cold War. Whether supplying intelligence information to analysts on the ground, aiding in battlefield communications, or accurately guiding weapons to their targets, they will continue to play an indispensable role in preserving the safety and security of the United States and the entire world.

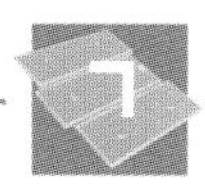

CHAPTER ONE

A COLD WAR IN SPACE

In 1945, the very idea of any sort of device being launched into space and placed in orbit around Earth fell somewhere between wishful thinking and science fiction. The technology required to loft a satellite into Earth's orbit was still undeveloped, although recent years had seen great progress toward that goal. A boom in research and development during World War II had led to the development of radar, improved radio communications, the first crude computers, the splitting of the atom, the jet engine, and, in the case of Nazi Germany, the first application of the rocket engine. All these new technologies would help contribute to realizing the dream of spaceflight.

OUT OF A NIGHTMARE, A VISION OF THE FUTURE

The German rockets, dubbed V-2s, were designed to be used as missiles against England during World War II. Capable of suborbital flight and guided by the crudest of navigation systems, the V-2 was the world's first ballistic missile. In the last year of the war, more than one thousand V-2s were launched at Great Britain, causing enormous devastation and loss of life. At a time when most of the world—and certainly all of Britain—saw the new rocket as a grim reaper of destruction, in the mind of a certain English scientist and science fiction writer, this powerful engine held the potential to benefit all of humankind.

From 1944 to 1945, over 3,000 German V-2 rockets were fired at London, Antwerp, and other Allied cities. Because the V-2 traveled at three times the speed of sound, it would land and detonate before the sound of its arrival could be heard by the residents.

Arthur C. Clarke was then an officer in Britain's Royal Air Force, as well as a member of Britain's Interplanetary Society. He would become a science fiction writer who would write such classic tales as *Childhood's End* and *2001: A Space Odyssey*. Clarke recognized the rocket's usefulness as a tool for peaceful scientific discovery. The February 1945 issue of *Wireless World*, a British technical journal, contained a letter from Clarke that first

Arthur C. Clarke has made a career of keeping his attention turned to the stars. In his book *Technology and the Future*, Clarke says, "The only way to discover the limits of the possible is to go beyond them into the impossible."

proposed the concept of "artificial satellites" in geosynchronous, or geostationary, orbit (see page 36) around Earth. Though Clarke's proposal centered on the postwar use of the V-2 rocket for upper atmospheric research, it also contained this additional suggestion:

"I would like to close by mentioning a possibility of the more remote future—perhaps half a century ahead. An 'artificial satellite' at the correct distance from the earth would . . . remain stationary above the same spot and would be within optical range of nearly half the earth's surface. Three repeater stations [satellites], 120 degrees apart in the correct orbit, could give television or microwave coverage to the entire planet."

As visionary as he was, Clarke could not have foreseen that it would take only twelve years for his dream of the artificial satellite to become a reality.

SPUTNIK

A Soviet tour guide holds up a replica of *Sputnik I* ("traveling companion"). The continuous beeping that *Sputnik I* broadcast during the three months it was in orbit was the first radio signal ever sent back to Earth from outer space.

Given the 1950s cold war mentality, when the United States and the Soviet Union were locked in a nuclear arms race and an ideological battle of wills, it is not surprising that the first satellite was launched not to benefit humanity but rather to intimidate a superpower adversary.

On October 4, 1957, the Soviet Union shocked the world with the launch from the Baikonur cosmodrome in Kazakhstan of *Sputnik I*, the first artificial satellite. The 184-pound (83.5-kilogram) satellite, less than two feet (0.6 meter) across, circled the globe once every ninety-six minutes and transmitted a steady beeping tone by radio signal to a receiving station on Earth. It was a humiliating public relations blow for the United States and called into question the country's assumption of technological superiority. The launching of *Sputnik I* also raised the genuine fear of the Soviets seizing

RUSSIAN SATELLITES

Though the United States has led the way in spy satellite technology and enjoys a clear advantage in space-based espionage efforts, Russia (part of the former Soviet Union) also has roughly one hundred military and civilian satellites currently in orbit. The fleet is aging, however, and "extremely insufficient," according to Russian general Anatoly Perminov (as quoted by the Associated Press). In 2001, Russian Aerospace Agency chief Yuri Koptev estimated that 80 percent of Russia's satellite fleet had already served longer than their designated lifetime. In recent years, Russia has not had enough money available to modernize its satellite systems, but it hopes to begin an intensive buildup in the very near future.

control of outer space and using it to gain military superiority over the rest of the world. It even seemed possible that the Soviets would soon be able to launch a network of orbiting weapons platforms from which they could rain nuclear devastation down upon the United States. The tiny, spinning, beeping piece of metal that was *Sputnik I* seemed to many like a grave threat to democracy.

THE SPACE RACE

In retrospect, *Sputnik I* provided a valuable service to the United States. The challenge it posed kicked the lagging U.S. space program into high gear. In answer to *Sputnik I*, the United States launched its first satellite, *Explorer I*, on January 31, 1958. From that point, the space race was on, and the United States would win most of the battles.

Technicians gather around *Explorer 1* as it is being installed on the Jupiter C rocket. Although it wasn't the first satellite in space, *Explorer 1* was responsible for discovering that large bands of radiation surround Earth—the first important discovery of the space age.

To the public, this "race" was exemplified by the United States's and the Soviet Union's manned space programs. The idea of manned spaceflight was given a clear focus and a concrete goal in 1961 when President John F. Kennedy challenged the country to land a man on the Moon and return him safely to Earth before the end of the

1960s. Kennedy's dream was realized in July 1969 when astronauts Neil Armstrong and Edwin "Buzz" Aldrin took the first steps on the Moon. Manned flight captured the public's imagination and put a human face on the race for technological dominance, but the real future of space exploration and development lay in unmanned spacecraft.

Throughout the 1960s, unmanned craft were being launched into the solar system by both nations to explore and photograph the Moon and the planets nearest Earth. But those who were skeptical of the value of both manned spaceflights and planetary exploration felt that the real space race was taking place behind the scenes. Deep within classified government agencies and top-secret laboratories, photoreconnaissance satellites were being developed that would enable nations to spy on their enemies from the safety of outer space.

THE U-2

President Dwight D. Eisenhower was a firm believer in the importance of photoreconnaissance (the photographing of enemy territory from above), knowing its tactical value from his experience as commander of the Allied Forces during World War II. Toward this end, the Central Intelligence Agency (CIA) tried using cameras mounted on unmanned balloons to spy on the Soviet Union, but these were too difficult to steer and control.

What was needed was a manned spy plane that could avoid detection while flying at very high altitude and photographing enemy territory. What was created was the U-2, a modified Lockheed F-104, capable of staying aloft for eleven hours, with a range of 4,750 miles (7,644 km). The U-2 flew over its target at altitudes of 70,000 (later 80,000) feet (21,336–24,384 m), far above Soviet air defenses and presumably undetected by their radar, snapping photographs with special cameras. The U-2 then returned to its base where the reconnaissance photos were developed and analyzed.

By the mid-1950s, the United States was flying regular U-2 missions over the Soviet Union, keeping a watchful eye on that nation's bomber and missile programs. The information that these flights provided gave the president an advantage in formulating foreign policy, knowing, for instance, that he could press for greater concessions from the Soviets because their military strength was not, in reality, as great as they claimed. The U-2 helped calm long-held fears of Soviet missile capabilities by revealing that the Soviet Union had yet to launch even an intercontinental ballistic missile (ICBM), a missile with a large enough range to reach targets in the United States.

The U-2 did not hold its tactical advantage for long, however. By 1958, the Soviets were not only tracking the planes by radar but also firing SAMs (surface-to-air missiles) at them. President Eisenhower was also concerned

Gary Francis Powers gazes at a model of the U-2 spy plane. Powers spent two years in a Soviet prison after the U-2 he was piloting was brought down over Soviet territory. Because he was not a soldier, he was not granted the rights given to a prisoner of war.

that the spy missions were causing tension in an otherwise improving relationship with Soviet premier Nikita Khrushchev. The president ordered the number of missions over the Soviet Union scaled back. Finally, with the approach in early May 1960 of a summit conference in Paris with France, England, and the Soviet Union, the spy flights were scheduled to be discontinued altogether.

The final U-2 flight, designated Operation Grand Slam, was scheduled for May 1, 1960, flown by veteran spy plane pilot Gary Francis Powers. As he made his way across the Soviet Union toward Plesetsk, a suspected

ICBM facility, his plane suffered a near miss with a Soviet SAM. The missile passed close enough to sever the plane's wings, necessitating Powers's ejecting from the plane and parachuting into Soviet territory.

Powers was arrested by the Soviets and the wreckage of his plane was recovered. Khrushchev used the incident to embarrass Eisenhower and the United States. Powers was held for two years before being released on February 10, 1962, in exchange for a Soviet masterspy, Colonel Rudolf Abel, who had set up a Soviet spy network in New York in the 1950s. Powers was criticized upon his return by some members of the CIA for not ensuring that the plane was destroyed or killing himself with poison to avoid interrogation. He died in 1977 at the age of forty-seven when a television news helicopter he was piloting crashed in Los Angeles.

More important than the renewed chill in U.S.-Soviet relations, the downing of the U-2 demonstrated the necessity of developing an "eye in the sky" that would be invulnerable to enemy attacks. The high-altitude spy camera was about to go even higher.

Fortunately, a program to replace the grounded U-2 was already in the works before Powers's crash. It would come to be called CORONA, though neither the program nor its name would become known to the public for more than thirty years.

CHAPTER TWO

CORONA

CORONA began in 1956 as an air force reconnaissance satellite program known as WS (Weapons System) 117L. To simplify and speed up the satellite's development, designers devised a system in which the exposed film would be returned to Earth in a reentry capsule rather than sending it back electronically (which would involve more intricate, expensive, and time-consuming design work). This was the program selected by President Eisenhower in the wake of the launching of *Sputnik I*. The CIA and the air force would share control of the program that would provide the United States with its first covert photoreconnaissance satellites.

A NEW ERA IN ESPIONAGE BEGINS

The result, after a dozen failed launches and lost rockets between February 1959 and August 1960, was *Discoverer XIV*, the first successful launch and deployment of a photoreconnaissance satellite and recovery of an ejected

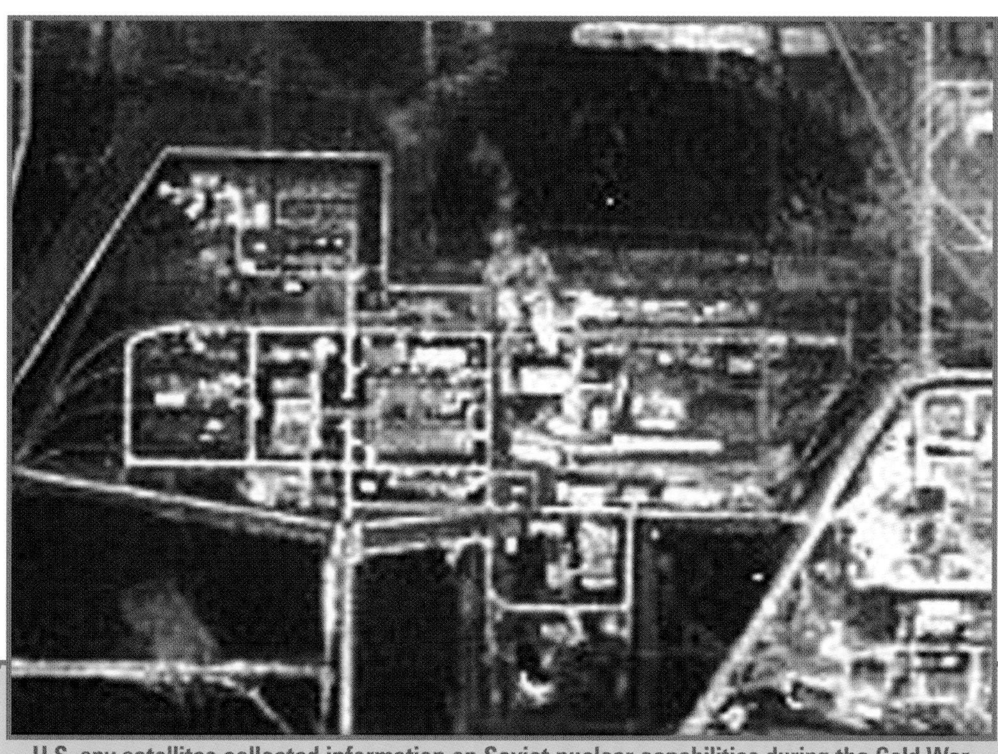

U.S. spy satellites collected information on Soviet nuclear capabilities during the Cold War in an attempt to determine the size of the USSR's nuclear stockpile. Facilities such as this one in Sverdlovsk, Russia, were once used to enrich uranium for use in nuclear weapons. Many are no longer operational now that the Cold War has ended.

film capsule. *Discoverer XIV* brought back images that revealed sixty-four new Soviet airfields and twenty-six new SAM sites. It also showed that the Soviet Union had only six ICBMs, rather than the previously estimated dozens. At a time when the two superpowers were engaged in a tug-of-war over the so-called missile gap (the difference between the number of missiles in the two country's respective nuclear arsenals), this intelligence came as a welcome relief to worried U.S. officials. It also had an effect

on how the United States conducted arms limitation talks with the Soviets, as well as on the number of new missiles the United States ordered to be built for its defense.

Similarly, in 1964, CORONA's cameras gave the United States advance warning that China was developing nuclear weapons even before any had been tested. This intelligence gave the United States a leg up on the diplomatic and propaganda fronts. More important, the United States's new spy satellites allowed it to monitor carefully and accurately Soviet and Chinese compliance with arms control treaties.

Before CORONA, the United States was entirely dependent on spies for intelligence gathering. But a spy, or even a whole network of spies, could not be everywhere, seeing everything, at all times. Without the ability to see exactly where enemy resources were located, how strong they were, and how or when they might be used, the United States could only piece together isolated parts of the whole picture. Spies could collect fragments of information, but even with thousands of them out in the field, they could still not find and assemble all the pieces of the puzzle.

In a nuclear age, the ability to see what the enemy was up to before it actually carried out its plans was desperately needed. The U-2 was the first step in the process of strengthening the United States's intelligence-gathering capabilities. CORONA—and the spy satellite programs

HIDDEN COSTS

Later model CORONA cameras were even more sensitive than earlier versions, but each development of the technology had its particular challenges to overcome. For instance, one reconnaissance satellite camera—previously available commercially and designed for use on Earth—required a modification of its shutter mechanism to work in the vacuum of space. Its lubricant had to be replaced with one that would function in the satellite's weightless environment. Engineers eventually determined the best lubricant for this job was whale oil—at a cost of $11,000 a gallon!

that have succeeded it—was the giant leap forward in technology that established and preserved the United States's dominance of space-based espionage.

CORONA'S CAMERA

In September 1961, the supersecret National Reconnaissance Office (NRO) was formed to oversee the design, building, and operations of America's spy satellites. Everything about these new "eyes in the sky" had to be created from scratch. The NRO continued a process of continuously developing and improving existing technologies that made for better, more reliable satellites capable of recording images of increasing resolution (clarity and detail). The earliest CORONA satellites used a twenty-four-inch (61-cm) panoramic camera developed by the Itek Corporation. A panoramic camera can capture

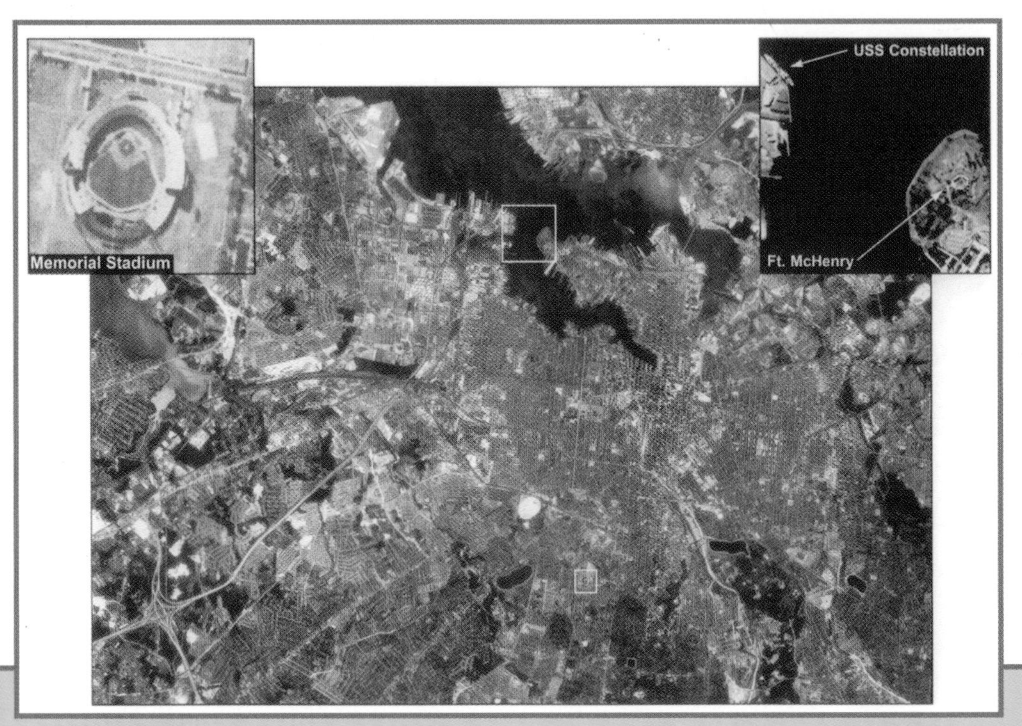

Although the first twelve CORONA satellite launches were failures and the thirteenth did not have a camera, the fourteenth took more photographs of the Soviet Union in one day than the U-2 spy plane took for the duration of its program. This CORONA photograph of Baltimore, Maryland, taken on September 25, 1967, shows the degree of detail that even primitive satellite photography achieved. The insets show Memorial Stadium and Fort McHenry.

an enormous area of land on one large, continuous photograph. The Itek camera could photograph an area ten miles by twenty-four miles (16 km by 38.6 km). It had a twenty-five-foot resolution, meaning objects twenty-five feet (7.6 m) or larger could be clearly identified in its photographs. The pictures were shot on a special fine-grain, seventy-millimeter film developed by the Eastman Kodak company for even greater clarity of the distant images.

The CORONA spy satellites needed a dependable way of getting their exposed film back to Earth. A small capsule

containing a large film take-up reel would, when full with film, separate from the satellite. After the satellite fired its own rocket to slow down by 1,300 feet (396 m) per second, gravity took hold and pulled the capsule back to Earth. A parachute opened at 50,000 feet (15,240 m), slowing the capsule's fall so that one of a fleet of Air Force C-119s, each equipped with a long nylon loop, could snatch it in midair. Failing a successful aerial recovery, helicopters and ships were also deployed in case the capsule fell into the sea.

BUILDING A BETTER SPY SATELLITE

While CORONA satellites were soon being launched at the rate of one per month, a constant program of design improvements was already in the works. Every phase of the program was under scrutiny, from launch vehicles to navigation systems, from the cameras to the film recovery capsules. Designers wanted to improve every step of the process to guarantee continued U.S. mastery of space-based espionage.

One problem was the short duration of CORONA missions, which were limited by the amount of film a satellite could physically carry into orbit. Lighter, more durable film stocks were developed, increasing the amount of film carried up to space and extending the life span of a mission. Some later CORONA satellites carried two film recovery capsules, allowing their cameras to keep clicking away while only one of the two capsules returned to Earth.

The designers' goal was to improve photo resolution, or the size of an object the cameras could capture with detailed clarity. Beginning at a twenty-five-foot resolution, CORONA camera systems improved to the point that a resolution of six feet (1.8 m) became possible. In the days since the end of the CORONA program, the resolution of spy satellites has been even more enhanced, although by how much remains classified information.

Over its twelve-year life span, until May 1972, the CORONA program flew 145 missions, with 120 of them considered complete or partial successes. These spacecraft sent back almost 400 miles (644 km) of film, containing more than 800,000 photographic images, covering some 557 million square miles (896 sq. km) of land area. As the program progressed into the 1960s, subsequent missions—all carrying the Keyhole, or KH designation—opened up Soviet military secrets to U.S. scrutiny. Thanks to satellite photoreconnaissance, U.S. military intelligence was able to locate and target virtually all the Soviet Union's ICBM, IRBM (intermediate-range ballistic missile), and ABM (antiballistic missile) sites, as well as all its secret warship, submarine, and military bases.

By the time CORONA conducted its final mission, satellite technology had evolved to the point where radio telemetry (the sending of information, including

photographs, by radio signal) would replace the film recovery capsule in the next generation of spy satellites as the fastest, most reliable method of transmitting satellite images to Earth. In its sixteen years, the program had gone through several generations of satellites that featured photographic, infrared, or radar devices designed to deliver improved reconnaissance images. These missions included the ARGON mapping satellites (1962–1964); LANYARD, designed to watch specific, fixed Soviet targets (1963); GAMBIT (1963–1984), which provided high-resolution images of smaller areas; and HEXAGON (1971–1984), with far greater resolution of larger areas. There were even plans, finally abandoned in 1969 after a billion dollars was spent on development, of a Manned Orbiting Laboratory (MOL) for the purpose of gathering electronic intelligence.

For almost two decades—during the height of the Cold War—CORONA kept its electronic eyes on the Soviet Union and its Communist allies and helped the United States navigate some of the scariest, most dangerous political waters in its history. But spy satellite technology, for all it had achieved until then, was still in its infancy. In the years to come, satellites would develop even further and really show the world—at least the part of it with the proper security clearance—what they could do.

CHAPTER THREE

INSIDE A SPY SATELLITE

In the world of technothrillers, technology is even more powerful and ingenious than it is in real life. On the movie screen or in the pages of a novel, spy satellites are routinely used to locate, identify, track, and even attack individuals. While these fictional satellites can read the headline off a newspaper or an address off a home from hundreds of miles up in space, the truth is somewhat less fantastic. Most experts do not believe real-life technology is up to the level of its fictionalized counterpart, even if it is, by the more conservative guesses, still impressive and awe inspiring.

Making educated guesses about the specifications of spy satellites is really about the best that can be done. Like the programs before them, the post-CORONA crop of orbiting reconnaissance satellites—known by code names such as KENNAN/CRYSTAL and LACROSSE—are top secret, their operations and missions classified. Based on published reports of commercial photoreconnaissance satellites and the informed opinions of experts, however, a reasonably clear picture of the current state-of-the-art spy satellites begins to emerge.

RESOLUTION

Most spy satellite–related topics remain classified, including the question of the resolution of the cameras aboard. Research and development of higher resolution cameras that could take more detailed pictures of ever-smaller objects continued for the duration of the CORONA project. Throughout the program's life, satellites were outfitted with the latest camera technology. Until the 1995 presidential executive order declassifying the CORONA project, this information continued to be top secret.

It was not generally known at the time that a six-foot (1.8-meter) resolution had been achieved by 1972, the year CORONA came to an end. Modern-day commercial photoreconnaissance satellites (which are privately owned and charge a fee for their imaging services) flown by companies such as EarthWatch and Space Imaging are known to have a one-meter (3.28-foot) resolution. Many experts suspect the resolution on U.S. spy satellites may not, at this point in time, be much greater.

Since 1994, the commercial resolution limit has stood at one-meter, indicating that a sub-single-meter resolution has most likely been achieved by the government's top-secret orbiters. In fact, some analysts at the

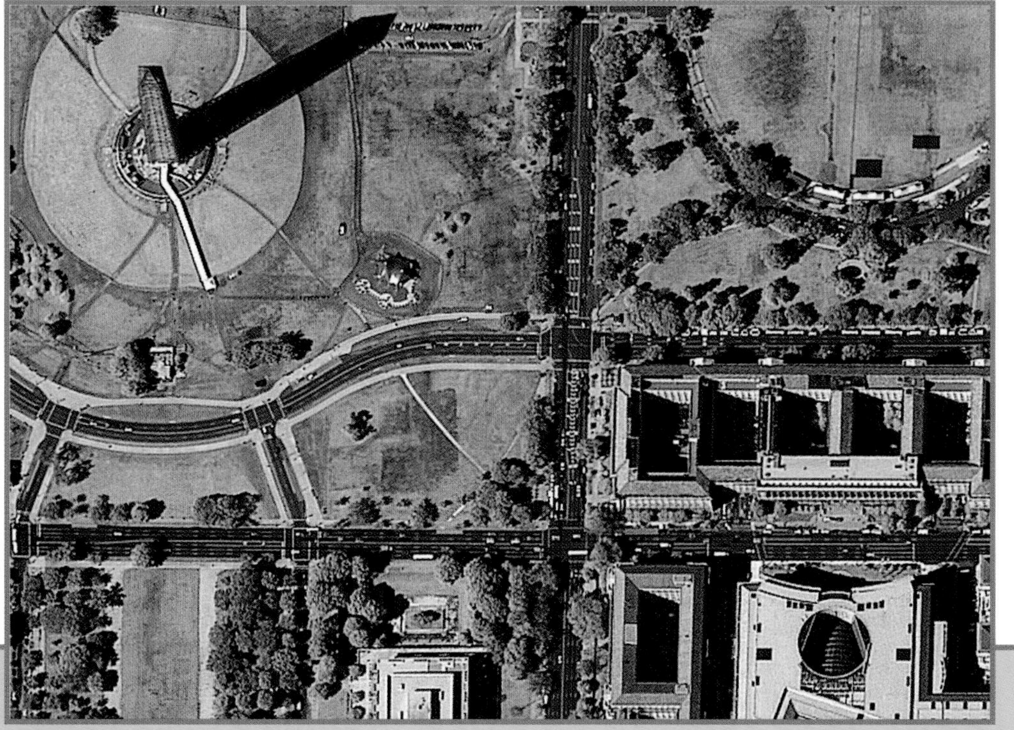

This detailed picture of Washington, D.C. was taken by the *IKONOS* satellite. Owned by a private company and orbiting the planet 14 times a day, *IKONOS* can detect objects as small as 3.3. feet (1 meter) across. By 2004, there will be an improved *IKONOS* that can detect objects as small as 19 inches (a half meter) across.

Federation of American Scientists (FAS) speculate that the NRO may have already achieved resolutions as small as 10 centimeters (3.9 inches). That is a high enough resolution to allow observers to see the baseball in the hand of a pitcher as a satellite passes some 290 miles (467 km) over a baseball diamond.

Since the earliest spy satellites started sending back their cargoes of exposed film, intelligence analysts have been faced with several obstacles in "reading"

the resulting pictures. Cameras had to deal with the distortion that came from shooting through Earth's atmosphere. Because Earth's atmosphere is thick with gases, light tends to bend and blur as it travels through space. There is also the problem of image smearing, which is an elongation of images caused by shooting from a great distance and at a slant range, or sharp angle. Image smearing made it difficult to obtain accurate height data on the objects the satellite was photographing (important information in determining the function of buildings or the size of a rocket on a launchpad). By having the cameras maintain a shallow and predetermined slant range, however, film analysts could determine the exact amount of distortion and compensate for it when studying the pictures.

Early on in the CORONA program, Itek Corporation developed the MURAL, or M camera system. The M system consisted of two cameras, one mounted tilted slightly forward and the second slightly backward, creating overlapping fields of vision. The two cameras would photograph the same swath of ground at the same fifteen-degree angle, but from different directions. When the two strips of film were aligned, the image would appear to be three-dimensional, giving a more accurate representation of the dimensions of the distant objects and helping to overcome the atmospheric distortions of space.

SPY SATELLITES: THE BASICS

The components of the current generation of spy satellites is classified information. Thanks to private companies such as EarthWatch and Space Imaging, however, informed guesses can be made based on what is known about commercial Earth imaging satellites built by companies such as Boeing Satellite Systems and Lockheed Martin Missile and Space. These companies, which also build classified spy satellites for the NRO, employ some of the same technology in their commercial satellites. Although their spy satellites seek out military targets, their commercial satellites are more likely to be used for more mundane tasks, like mapping terrain, recording environmental observations, and taking geological readings for use in mining.

In general, all satellites feature the same basic components, or subsystems, although designs and functions vary depending on the task for which they were designed. These include:

- Command and Data-Handling Subsystem: This serves as the satellite's brain, controlling the spacecraft's functions through the flight computer. All the commands that tell the satellite what to do and when to do it are relayed through this subsystem. It also handles all collection, processing, and storing of data.

- The Power System: Satellites generally feature solar panels that gather sunlight and store it as energy in batteries that distribute the power to the rest of the instruments. These panels can generate more than 1,500 watts of power.

- Attitude Control Subsystem: This helps to maintain the satellite in its orbit, keeping it steady and pointed in the proper direction. The subsystem uses sensors keyed to celestial objects, such as stars, or infrared detectors to gather precise positioning data. An onboard propulsion system will then receive instructions from Attitude Control to fire small cold gas rockets to correct or change the satellite's orbit if necessary.

- The Payload: These are the instruments a satellite uses to carry out its mission. This can range from antennas for the transmission of telephone or TV signals to reconnaissance cameras to more sophisticated imaging or data retrieval equipment.

- Communications Subsystem: This allows the satellite to receive commands from ground controllers and relay the information it has gathered down to Earth.

- Control Subsystem: Heat distribution units and thermal blankets are used in combination to protect the spacecraft's delicate electronics from the damage associated with the extreme temperature swings of space. While in orbit, satellites pass regularly into and out of sunlight, exposing them to temperatures ranging from approximately -190° F (-87.8° C) in the shade to 350° F (176.7° C) in direct sunlight.

- Ground System: A satellite ground system is a station or a series of stations on Earth that both receives a satellite's transmissions and sends instructions or commands up to the satellite. The system can include a spacecraft control center, ground stations to send commands and receive data, a data-handling facility, and a data archive.

ASSEMBLY

Although we do not know precisely what components go into a spy satellite or exactly how they are produced, the manufacturing and launching processes must be very similar to those of other satellites. Boeing Satellite Systems (BSS) of El Segundo, California, manufactures satellites for use by private companies, launching an average of one satellite per month. These are all designed, manufactured, and shipped from their Integration and Test Complex—a former American Motors Corporation factory where the Nash Rambler automobile used to be manufactured—an

840,000 square foot state-of-the-art facility.

BSS uses an assembly line process to construct its satellites. The Integration and Test Complex is divided into a series of bays—areas of specialization—that the satellite passes through on its way to completion, much like the way a car is constructed piece by piece as it passes down the assembly line. The body of the satellite is built first in the structure bay, after which it passes through the propulsion bay where its maneuvering rockets and boosters are inserted. Its next stop is payload integration and testing, where the specialized equipment that the satellite was created to carry into orbit is added and tested to ensure that it is functioning as intended by the designers. After that, the satellite is rolled into the antenna fabrication bay, where its communications antennae are assembled and attached. These allow the satellite to send and receive information, such as photos,

Boeing Satellite Systems, headquartered in El Segundo, California, is one of the largest satellite manufacturers in the world.

SECRET NO MORE

Since the beginning of the CORONA program and the failed launch of its first satellite in 1959, the NRO had kept all spy satellite launches top secret. Though the public was kept in the dark and all information relating to the satellites was classified, many aerospace experts were aware of these launches and had at least some idea of what each new satellite could do. In December 1996, however, the NRO broke with almost forty years of Cold War policy and announced a spy satellite launch ahead of time. Although no details on the satellite were released, many people believe it was a LACROSSE radar imagery satellite that can take high-resolution pictures even through heavy cloud or fog cover.

cell phone calls, and radio and television signals. Next, the satellite is rolled into the solar array assembly bay. At this point, its power-generating solar panels are attached.

The last stop before the finished satellite can be shipped to the launch site is the high bay, where it undergoes final integration and a rigorous series of tests. The satellite then enters the Mass Properties Laboratory where it is weighed and checked for "spin balance"—a test of its center of gravity, which ensures the satellite will maintain a smooth, even orbit in the weightlessness of space.

Next, in the Space Simulation Laboratory, the satellite undergoes a series of thermal, vibration, and shock tests. A "vibration table" shakes the spacecraft, vibrating from different directions and with varying intensity to

make sure the satellite can withstand the rigors of launch and orbit. A thermal vacuum chamber duplicates the temperature and atmospheric conditions the satellite will encounter in orbit. Here, it is subjected to a vacuum (a space empty of all matter) like the one in which it will operate in space, as well as to the wide range of temperatures it will be exposed to as it circles Earth, moving from direct exposure to the sun to deep darkness.

A Titan 4B rocket lifts off from Vandenberg Air Force Base in California, carrying a classified NRO satellite into orbit. Vandenberg AFB is the only military base in the United States from which unmanned government and commercial satellites are launched into polar orbit.

LAUNCH

Once these exhaustive tests have been completed, the satellite is packed in a shipping container for transportation to the launch site. There, it is either mated to a rocket or prepared for inclusion in the cargo hold of the

space shuttle—the two vehicles that can deliver satellites into orbit. The type of orbit it will follow once launched beyond Earth's atmosphere depends on its specific tasks. Possible orbits include:

- Geosynchronous Equatorial Orbit (GEO): This places the spacecraft directly over the equator in an east-west direction at an altitude of 22,300 miles (35,888 km). At that distance, the satellite and Earth take exactly the same amount of time to rotate around Earth's axis, so the satellite remains fixed over the same spot on the globe. For this reason, it appears not to be moving at all. This type of orbit is most commonly used for communications satellites.

- Low Earth Orbit (LEO): This places the satellite between 200 and 500 miles (322 to 805 km) above the planet, again travelling in an east-west direction. This lower orbit offers a satellite's cameras or imaging systems a closer look at Earth's surface. At that distance, the spacecraft circles Earth once every ninety minutes at approximately 17,000 miles per hour (27,359 km/hr), a speed that prevents gravity from pulling the satellite back into the atmosphere. The space shuttle and the international space station travel in LEO.

- Polar Orbit: This is a type of LEO, but the satellite circles Earth in a north-south direction—over the poles—rather than in an east-west direction over the equator. Because a polar orbiting satellite eventually covers much of the earth's surface, it is useful for monitoring the global environment.

- Elliptical Orbit: This follows an oval-shaped, north-south path and takes about twelve hours to circle the planet. In an elliptical orbit, also known as a reconnaissance orbit, the satellite does not remain a fixed distance above Earth; sometimes the orbit brings it closer, other times it takes it farther away. This approach-and-withdrawal orbit has the advantage of allowing a satellite to bring its cameras in close to get a detailed look at a target, as well as to pull back and broaden its footprint (the amount of surface area falling under its view).

Whether in an orbit near or far, the new spy satellite will settle into its path, there to unfurl its electronic eyes or ears and begin the work for which it was designed.

CHAPTER FOUR

EYES IN THE SKY

The United States began flying photoreconnaissance satellites in 1959 under two separate programs, the U.S. Air Force's SAMOS and the CIA's CORONA. Both of these programs suffered delays and setbacks as a result of technical problems, however. So President Eisenhower ordered a study that lead to, first, the creation of the Office of Missile and Satellite Systems, and, later, in September 1961, to the formation of the National Reconnaissance Office (NRO), a combined U.S. Air Force and CIA operation.

THE NRO

The NRO was established as a covert (top secret) organization, headed by the CIA's deputy director for plans and overseen by the Department of Defense and the air force. It remained top secret for more than three decades, until the organization—but not the details of its current operations—was declassified in 1995. It took two more years to reveal the location of its headquarters and another year after that for the more than 800,000 photoreconnaissance images taken between 1960 and 1972 to be made public.

Located in Chantilly, Virginia, the NRO's stated mission is to be "Freedom's Sentinel in Space: One Team, Revolutionizing Global Reconnaissance." Their satellites warn of political trouble spots, monitor the global environment, and plan military strikes.

The NRO is charged with designing, building, and operating space reconnaissance systems that will allow the United States to maintain its edge in space-based information gathering. The research and development that goes into creating and maintaining a system of spy satellites takes place under the tightest security. Its "customers"—the groups that use the information its satellites gather—include the CIA and the Department of Defense (DOD). NRO satellites can warn of potential trouble spots around the world, help plan military operations, and monitor changes to the environment. As a DOD agency, the NRO is staffed by DOD and CIA personnel and is funded through the National

THE SATELLITE WAR ON TERROR

It is almost certain that NRO-operated satellites recorded images of the September 11 attacks, though none have been released to the public. Far more important than supplying alternate aerial views of this tragedy, however, is the role spy satellites have been playing ever since.

Within a week of the attack, existing signals intelligence (sigint) satellites—which intercept radio, cell phone, and other electronic signals—were believed to have been "retasked," or reprogrammed, to turn their electronic ears on Afghanistan and aid in the hunt for Osama bin Laden and members of Al Qaeda, the group thought to be responsible for the September 11 attacks. It was hoped that the terrorists would betray their location or their plans by sending out detectable cell phone and e-mail communications. Two other commercial spy satellites designed to take high-resolution images from space (and owned and operated by private companies that sell their images to the public) were also believed to have been pressed into duty by the government and retasked for war-time duty over Afghanistan.

Reconnaissance Program, part of the National Foreign Intelligence Program.

How the NRO achieves its goals, and exactly how big a budget that requires—currently believed to stand at more than $6 billion a year—remains classified information, even if its existence is no longer a secret. Headquartered in Chantilly, Virginia, the NRO employs approximately three thousand people in its quest for "global information superiority." But for all its secrecy, Dwayne Day, a Washington, D.C., researcher who has

written about spy satellites, told Space.com: "[The NRO] probably ranks as one of the most important government agencies during the Cold War . . . [when] it helped to stabilize the U.S. We didn't experience so many surprises anymore."

Keeping military and political surprises to a minimum is the aim of the NRO. In the wake of the September 11 attacks, this has become more important than ever. Coincidentally, a series of launches begun on August 17, 2001, to replace old satellites and insert new ones into orbit was already underway when Al Qaeda launched its deadly attacks on the Pentagon near Washington, D.C., and the World Trade Center in New York City. These new spy satellites will vastly improve and increase U.S. reconnaissance capabilities, though it remains to be seen how a high-tech eye in the sky can guard against low-tech terrorist methods, like box cutter–wielding hijackers or individuals standing in a crowded shopping center with small but powerful nail bombs hidden under their jackets.

OTHER SPY SATELLITE SYSTEMS

Photoreconnaissance is far from the only type of imaging satellite circling above us. Standard optical-system satellites use a large mirror to gather light to create photographic images (imagine the Hubble Telescope, only with

In downtown Jerusalem, Israeli police secure the area following a suicide bombing. Thirty people were injured when a woman detonated explosives concealed under her clothing on a crowded street. No matter how sophisticated and technologically advanced they become, spy satellites may never be able to alert us to and protect us from this kind of terrorist attack.

its lens pointed down at Earth rather than off into deep space). Radar-imaging satellites use microwave signals to peer through Earth's atmosphere (even penetrating cloud cover) to scan the planet's surface. Combination radar and optical satellites are also in use, offering wider areas of view in sharper detail than either could offer separately. Signal intercept and detection satellites pick up enemy radio, telephone, and data transmissions, and ocean observation satellites are used to locate and track ships at sea.

The United States today fields a wide variety of spy satellites, including:

- KH-12 (Improved CRYSTAL): First launched in November 1992. The KH-12 features electronics that allow for sharper images and higher resolutions. It features a periscope-like rotating mirror that reflects images onto a primary mirror. Because the rotating mirror can be sharply angled, it allows the satellite to photograph objects hundreds of miles away from its flight path.

- INDIGO/LACROSSE/VEGA: A series of radar-imaging satellites first launched in January 1982 from the space shuttle. By using radar instead of light to image its target, these satellites are able to "see" through clouds and fog with a resolution believed to be in the one-meter range. Its images are thought to rival those of photoreconnaissance satellites. Because this is an ongoing spy satellite program, however, details remain classified.

- MERCURY/INTRUDER/PROWLER: Signals intelligence (sigint) satellites were first launched in the 1990s and designed to intercept enemy broadcast, communications, radar, telemetry, and other electronic signals.

- WHITE CLOUD: A naval ocean surveillance system (NOSS), first launched in 1986 and created to keep its electronic ears on wide areas of the ocean, determining the location of radio and radar signals emanating from ships at sea. NOSS is believed to consist of a cluster of one main satellite and three smaller subsatellites that trail along behind it in a low elliptical orbit.
- RANGER: First launched in 1990, RANGER is a naval program believed to be involved in signal and electronic intercepts. It is also speculated that, like WHITE CLOUD, RANGER consists of a primary satellite trailed by three subsatellites.
- The Defense Meteorological Satellite Program (DMSP): These satellites, first launched in 1992, monitor weather conditions in target areas as an aid to military planners in determining the timing of troop deployments and movements and air strikes.

COMMERCIAL USES FOR SPY SATELLITE TECHNOLOGY

For all their importance as information-gathering tools during last century's Cold War and the new century's

war against terrorism, spacecraft that feature spy satellite technology have been applied to a wide range of other, less secretive tasks, ranging from geological surveys and mapmaking to storm tracking, law enforcement, and search and rescue missions.

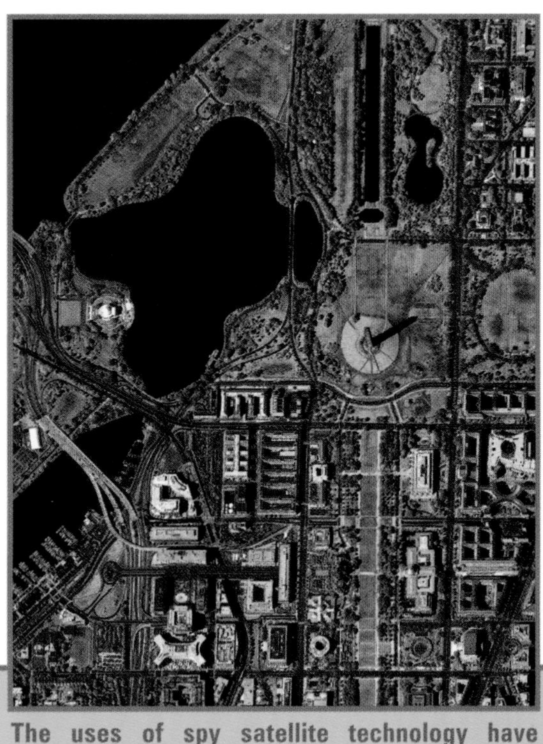

The uses of spy satellite technology have expanded over the decades, making it possible for cartographers, geologists, and urban planners to also benefit from the precise images that high-resolution aerial photographs provide.

For instance, data from reconnaissance satellites can be used for exploration and mapping. Maps featuring minute detail—including features as small as foot trails or streams meandering through wooded areas—can be created based on high-altitude reconnaissance photographs.

Satellites offer the ability to track and predict emerging weather patterns. This allows for long-range forecasting that allows communities to plan for droughts, severe winters, or the flooding associated with rainy springs and melting snowpack. Satellites also help meteorologists develop very accurate short-term forecasts, enabling

COMPETITION IN SPACE

Though it is believed currently to have almost one hundred orbiting spacecraft working in support of military and intelligence activities, the United States no longer enjoys exclusive control of space. Once, only the Soviet Union (and later Russia) had satellite technology that rivaled that of the United States. Now, however, the governments of Japan, Canada, Israel, France, China, and several other nations also maintain their own satellites, as do several private U.S. and foreign firms. They are not as good as those operated by the NRO, but these private satellites can still provide solid military intelligence to foreign governments and even paying customers.

Operators of commercial satellites now sell their high-resolution images to anyone who can afford them (for between $3 and $7 per square kilometer of photographed territory). It is conceivable that enemy countries or terrorist organizations would be able to purchase images to show them U.S. troop movements, as well as to track fleets, count tanks, and gain other information that could tip them off to U.S. military plans. They could also gain access to highly detailed images of U.S. cities, military installations, government buildings, and other tempting targets of attack. Despite the State Department's right to shut down commercial imaging satellites during a crisis, critics of commercial spy satellites remain alarmed. As they point out, terrorists and dictators will buy their images before they launch an attack and set off a crisis, not after.

According to CBSNews.com, in 2002, CIA director George Tenet told the Senate, "The unique space-borne advantage that the U.S. has enjoyed over the past few decades is eroding . . . Foreign military, intelligence, and terrorist organizations are exploiting this—along with commercially available navigation and communications services—to enhance the planning and conduct of their operations." James Lewis, of the Center for Strategic and International Studies, agreed: "We're losing our monopoly."

authorities to issue early warnings to those living in the path of a hurricane or a tornado. In agriculture, satellites can provide detailed mapping and surveying of crops, giving farmers far more accurate tools with which to predict how large and healthy their crops are and how much they are likely to get at market.

Satellites can also help save the day in emergencies. For example, they can locate the source and severity of forest fires and aid emergency workers in determining which resources are needed to fight them and where they should be placed. Downed airplanes or ships in distress can be located from space, helping rescue and recovery crews get to accident sites as quickly as possible. Satellite images of areas flooded by swollen rivers can help rescue and salvage crews target their efforts where they are most needed. This satellite information can provide the evidence necessary for a community to receive federal disaster relief funds.

While spy satellites were designed with military applications in mind, the technology that has gone into them allows for more efficient coordination of emergency relief and rescue efforts. In this way, spy satellite technology can help us develop stronger bonds of friendship and cooperation throughout the world.

CONCLUSION

The world is a very different place today than it was when the first spy satellites were launched in the late 1950s. Then, the United States was in the middle of the Cold War with the Soviet Union and its Communist allies—a conflict that featured clearly defined enemies contained within specific borders and backed by an arsenal of known weapons and resources.

As the events of September 11 made abundantly clear, however, enemies of the United States are no longer as easily identifiable, nor are they neatly confined within clearly drawn national borders. Yet the country's present system of spy satellites was designed to watch over yesterday's foes.

In order to meet the demands of the future and satisfy the widening range of users of satellite-gathered intelligence, entirely new systems are needed. From the ability to provide higher resolutions to real-time surveillance (hovering over a fixed location and providing a continuous live feed to controllers on Earth), to creating complex, multimedia data (combining several different sources of data, such as photographs taken on the ground, technical information, satellite images, and three-dimensional computer modeling to form a single, comprehensive picture of a location), the next generation of satellites needs to be developed and launched as soon as possible.

To help meet that need, the NRO is developing the Future Imaging Architecture (FIA). Based on current small satellite technology, FIA is the imaging intelligence technology expected to operate for the next several decades. It will deliver more data at a faster rate and with higher resolution, with as much as twenty times the capability of current satellites. Begun in 1998, the first FIA-based satellite—to be created as part of the estimated $4.5 billion program, the most expensive in the history of the intelligence community—is expected to be launched no later than 2005.

The Boeing Company is building this next generation of satellites for the NRO. As part of this effort, Boeing bought, of all things, a former movie studio special effects company, Autometric of Springfield, Virginia. Autometric owns a three-dimensional graphics program called the Enhanced Digital Geodata Environment (EDGE), which converts images into advanced graphic simulations. Soon, spy satellites may be able not only to provide high-resolution photographs but also to create computer-generated, three-dimensional images of locations based on those "flat" photographs. This will allow analysts at their computers to enter and travel through a location as though they were actually there, getting an accurate feel for its terrain, streets,

Run by the NSA with help from agencies in England, Canada, Australia, and New Zealand, ECHELON is a series of satellites and earth stations designed to intercept virtually every single phone call, fax, and e-mail message sent anywhere in the world. The NSA processes all the information that is intercepted, searching for certain words and phrases that might indicate a security threat.

or layout. Troops or special operations forces, for example, will be able to use these images to learn their route and its dangers before they are deployed, making for a safer and more efficient operation.

FIA is also expected to include a space radar surveillance system, known as Starlite, aimed at creating all-weather, broad area, and near continuous radar access to assist in military operations. There have also been calls for so-called real-time satellite sensor and surveillance

systems, providing the intelligence, military, and law enforcement communities with a see-it-as-it-is-happening view of the world.

Electronic eavesdropping is already a fact of life under programs such as ECHELON (a joint effort of the United States, Canada, Australia, New Zealand, and England), which use a network of land-based and satellite listening posts to keep tabs on radio and microwave transmissions. Eventually, this technology will improve so that literally hundreds of millions of daily telephone conversations, faxes, and e-mail messages will be routinely scanned by computers programmed to "listen" for key words that might indicate a security threat.

It is expected that private photoreconnaissance satellites will also play an increasingly large role in national security. Infrared and spectrographic camera technologies are being developed to assist in the tracking of troop movements, the monitoring of ship and air traffic, and the testing of the atmosphere for pollution or traces of radioactive, chemical, or biological weapons.

Ever since Arthur C. Clarke's proposal for a geosynchronous artificial satellite, aerospace engineers have dreamed of creating a satellite system that would cover the entire surface of Earth, so that at any given time, every inch of land and sea was under the watchful "eye" of a satellite. We may soon be one step closer to realizing that dream. In May 1999, a LACROSSE series satellite was

launched, believed to be the first in a series of twenty-four multifunction satellites that will eventually circle the globe and provide a tight web of reconnaissance coverage—any given spot on the planet will be passed over by a LACROSSE satellite every fifteen minutes.

The greater watchfulness that these LACROSSE satellites and developing reconnaissance technology will provide cannot come at a better time. September 11 showed us that as good as our reconnaissance networks are, there are still gaping holes in their coverage. The next generation of satellites will help to fill those holes and, it is hoped, prevent a similar surprise attack. By knowing what enemies of the United States are planning before they strike, both lives and property can be protected. Only by watching our enemies from the high ground of outer space can we keep one step ahead of those who would deny us and the rest of the world the peace, safety, and security that is every person's right.

GLOSSARY

antiballistic missile (ABM) A missile designed to shoot down other missiles.

Cold War The period between the end of World War II and the 1991 dissolution of the Soviet Union, referring to the political tension and military rivalry between the United States and the USSR that stopped short of actual full-scale war.

command and data system The satellite's brain, controlling the spacecraft's functions through the flight computer.

COMSAT Communications satellite.

CORONA The United States's first spy satellite program, which launched 145 missions between 1960 and 1972.

elliptical orbit An oval-shaped, north-south orbit that takes about twelve hours to circle Earth. Also known as a reconnaissance orbit.

geosynchronous or geostationary equatorial orbit (GEO) An orbit that places a spacecraft directly over the equator in an east-west direction at an altitude of 22,300 miles (35,888 km). Since the satellite travels at the same speed as Earth does spinning on its axis, the satellite remains fixed over the same spot on the globe.

global positioning system (GPS) Satellite-based radio positioning/navigation system that provides extremely accurate, three-dimensional, common grid position, velocity, and time-of-day information to users anywhere on or near Earth.

HUMINT Human intelligence, or intelligence information gathered by human agents rather than machines such as satellites.

intercontinental ballistic missile (ICBM) A missile capable of carrying a payload a long distance, from one continent to another.

intermediate-range ballistic missile (IRBM) A missile capable of carrying a payload a medium distance.

keyhole Designation for any one of a series of photoreconnaissance satellites that uses an orbital telescope for photoreconnaissance.

low earth orbit (LEO) An orbit between 200 and 500 miles (322 and 805 km) above the planet, travelling in an east-west direction.

Naval Ocean Surveillance System (NOSS) A satellite designed to keep its electronic ears on wide areas of the ocean, determining the location of radio and radar signals emanating from ships at sea.

National Reconnaissance Office (NRO) A joint Department of Defense/CIA agency responsible for the engineering, development, acquisition, and operation of space reconnaissance systems.

optical-system satellites A satellite that uses a large mirror to gather light to create photographic images.

orbit The path of a celestial or human-made satellite as it revolves around another body.

payload The cargo, equipment, or explosive charge carried by a satellite or missile.

photoreconnaissance The surveying of a military target through aerial or satellite photography.

pointing control The onboard navigation and propulsion system that maintains the satellite in orbit, keeping it steady and pointed in the proper direction.

polar orbit A type of LEO that parks the satellite in an orbit that circles Earth in north-south direction—over the poles—rather than in an east-west, or equatorial, direction.

radar-imaging satellites Satellites that use microwave signals to "see" through cloud cover to scan the Earth's surface.

resolution The degree of clarity of an image.

surface-to-air missile (SAM) A missile fired from a ground-based launcher at an airborne target.

SIGINT Signal intelligence, or the gathering of information through the detection and interception of radio, telephone, microwave, and data transmissions.

telemetry The transmission of data by radio or other means from a remote source to a receiving station for recording and analysis.

USSR Union of Soviet Socialist Republics. The name of the confederacy of Russia and her satellite republics prior to its 1991 breakup.

wireless The British term for radio.

Boeing Satellite Systems
Boeing Military Aircraft and Missile Systems
P. O. Box 516
St. Louis, MO 63166
(314) 232-0232
Web site: http://boeing.com/defense-space/space/sitemap.html

Central Intelligence Agency (CIA)
Office of Public Affairs
Washington, D.C. 20505
Web site: http://www.cia.gov/cia/ciakids/index.html

Department of Defense
Office of the Secretary of Defense
Public Affairs
OASD(PA)/PIA
1400 Defense Pentagon, Room 3A750
Washington, DC 20301-1400
(703) 428-0711
Web site: http://www.defenselink.mil

Jet Propulsion Laboratory
Public Services Office
Mail Stop 186-113

4800 Oak Grove Drive
Pasadena, CA 91109
(818) 354-0112
Web site: http://www.jpl.nasa.gov

NASA Headquarters
Information Center
Washington, DC 20546-0001
(202) 358-0000
Web site: http://www.nasa.gov

National Reconnaissance Office
14675 Lee Road
Chantilly, VA 20151-1715
Web site: http://www.nro.gov

U.S. Strategic Command
Public Affairs Office
901 Sac Boulevard, Suite 1A1
OFFUTT Air Force Base, NE 68113-6020
(402) 294-4130
Web site: http://www.stratcom.mil

WEB SITES

Due to the changing nature of Internet links, the Rosen Publishing Group, Inc., has developed an online list of Web sites related to the subject of this book. This site is updated regularly. Please use this link to access the list:

http://www.rosenlinks.com/ls/spsa/

FOR FURTHER READING

Becklake, Sue. *Space, Stars, Planets, and Spacecraft.* New York: Dorling Kindersley, 1998.

Branley, Franklyn Mansfield. *From Sputnik to Space Shuttles: Into the New Space Age*. New York: HarperCollins, 1989.

Gaffney, Timothy R. *Secret Spy Satellites: America's Eyes in Space.* Berkeley Heights, NJ: Enslow Publishers, 2000.

Graham, Ian. *Planes, Rockets, and Other Flying Machines.* New York: Franklin Watts, 2000.

Graham, Ian. *Satellites and Communications.* Austin, TX: Raintree/Steck Vaughn, 2000.

Kallen, Stuart A. *The Race to Space*. Edina, MN: ABDO Publishing, 1996.

Mellett, Peter. *Launching a Satellite*. Crystal Lake, IL: Heineman Library, 1999.

Stott, Carole. *Space Exploration*. London, England: Dorling Kindersley, 1997.

Vogt, Gregory. *Rockets (Exploring Space)*. Bloomington, MN: Bridgestone Books, 1999.

Walker, Niki. *Satellites and Space Probes.* New York: Crabtree Publishers, 1998.

BIBLIOGRAPHY

Andronov, Major A. "American Geosynchronous SIGINT Satellites." FAS.org. 1993. Retrieved February 2002 (http://fas.org/spp/military/program/sigint/androart.html).

Associated Press. "Russia Plans to Increase Spy Satellites to Keep Watch on Afghanistan." Space.com. October 3, 2001. Retrieved May 2002 (http://www.space.com/news/russia_sats_011003.html).

Berner, Steve. "Fighting Proliferation: Chapter 5, Proliferation of Satellite Imaging Capabilities: Developments and Implications." FAS.org. January 1996. Retrieved February 2002 (http://www.fas.org/irp/threat/fp/index.html).

"Chronology of Spy Satellites." Infomanage.com. 1998. Retrieved February 2002 (http://infomanage.com/international/intelligence/spychron.html).

Clarke, Arthur C. *Ascent to Orbit: A Scientific Autobiography*. New York: John Wiley & Sons, 1984.

Cosmiverse Staff Writer. "Spy Satellites Search for Terrorists." Cosmiverse.com. September 19, 2001. Retrieved April 2002 (http://www.cosmiverse.com/space09190104.html).

Curtis, Anthony R. *Space Satellite Handbook*. Houston, TX: Gulf Publishing, 1994.

David, Leonard. "The NRO: Dark Secrets Under Open Skies." Space.com. September 26, 2000. Retrieved February 2002 (http://www.space.com/news/spacehistory/nro_first_side_000926.html).

Day, Dwayne A., John M. Logsdon, and Brian Layell, eds. *Eye in the Sky: The Story of the Corona Spy Satellites*. Washington, DC: Smithsonian Institution Press, 1998.

Defense Science Board. "Defense Science Board Task Force on Satellite Reconnaissance." FAS.org. January 1998. Retrieved April 2002 (http://www.fas.org/spp/military/program/imint/dsb-980100.html).

Halberstam, David. *The Fifties*. New York: Fawcett Columbine Books, 1993.

Hartenstein, Bill, and the Associated Press. "Titan 4 Launches U.S. Spy Satellite." Florida Today Space Online. December 20, 1996. Retrieved May 2002 (http://www.floridatoday.com/space/explore/stories/1996b/122096e.htm).

Havill, Adrian. *The Spy Who Stayed Out in the Cold*. New York: St. Martins Press, 2001.

"High Tech Spy Satellites Not Targeting Americans, CIA, NSA Directors Say." CNN.Com. April 12, 2000. Retrieved May 2002 (http://www.cnn.com/2000/US/04/12/spies.speak).

Lumpkin, John J. "Satellites: Everyone's Got 'Em." CBSNews.Com. April 9, 2002. Retrieved May 2002 (http://www.cbsnews.com/stories/2002/04/09/tech/main505705.shtml).

Maral, Gerard, and Michel Bousquet. *Satellite Communications Systems*: *Systems, Techniques and Technology*. New York: John Wiley & Sons, 2002.

Montenbruck, Oliver, and Eberhard Gill. *Satellite Orbits: Models, Methods, Applications*. New York: Springer Verlag, 2000.

Peebles, Curtis. *The CORONA Project: America's First Spy Satellites*. Annapolis, MD: United States Naval Institute Press, 1997.

Richelson, Jeffrey T. *America's Secret Eyes in Space: The U.S. Keyhole Satellite Program*. Cambridge, England: Ballinger Publishing, 1990.

"Satellite, Space Station Crew Watch Horror." CNN.com. September 13, 2001. Retrieved June 2002. (http://www.cnn.com/2001/us/09/12/satellite.images).

Windrem, Robert. "Spy Satellites Enter New Dimension." MSNBC.Com. August 8, 1998. Retrieved May 2002 (http://www.msnbc.com/news/185953.asp#BODY).

INDEX

ABOUT THE AUTHOR

Paul Kupperberg is a freelance writer and an editor for DC Comics. He has published more than seven hundred comic book stories, books, and articles, as well as several years of the syndicated *Superman* and *Tom and Jerry* newspaper strips. *Spy Satellites* is his second book for Rosen Publishing. Paul lives in Connecticut with his wife, Robin, and his son, Max.

PHOTO CREDITS

Cover, p. 6 © Reuters NewMedia, Inc./Corbis; p. 9 © Corbis; pp. 10, 11 © Bettmann/Corbis; p. 13 © Marshall Space Flight Center/NASA; pp. 16, 19, 33, 42 © AP/ Wide World Photo; p. 22 © MAI/TimePix; p. 28 © Space Imaging/AP/Wide World Photo; p. 39 © Greg Mathieson/MAI/Timepix; p. 45 © AFP/Corbis; p. 50 © Michael Dalder/Reuters/TimePix.

DESIGNER

Tom Forget